A giant porthole, vista onto another world, from Jules Verne's Twenty Thousand Leagues Under the Sea, *1870*

Bloomsbury USA

An imprint of Bloomsbury Publishing Plc

1385 Broadway	50 Bedford Square
New York	London
NY 10018	WC1B 3DP
USA	UK

www.bloomsbury.com

BLOOMSBURY and the Diana logo are trademarks of
Bloomsbury Publishing Plc

First U.S. edition 2018

© Philippa Lewis, 2018

All rights reserved. No part of this publication may be reproduced or
transmitted in any form or by any means, electronic or mechanical,
including photocopying, recording, or any information storage or retrieval
system, without prior permission in writing from the publishers.

No responsibility for loss caused to any individual or organization acting on
or refraining from action as a result of the material in this publication can be
accepted by Bloomsbury or the author.

ISBN: HB: 978-1-63557-084-7

Library of Congress Cataloging-in-Publication Data is available.

2 4 6 8 10 9 7 5 3 1

Designed and typeset by Wooden Books Ltd, Glastonbury, UK

Printed in the U.S.A. by Worzalla, Stevens Point, Wisconsin

To find out more about our authors and books visit
www.bloomsbury.com. Here you will find extracts, author interviews,
details of forthcoming events, and the option to sign up for our newsletters.

Bloomsbury books may be purchased for business or promotional use.
For information on bulk purchases please contact Macmillan Corporate and
Premium Sales Department at specialmarkets@macmillan.com.

PORTALS

GATES, STILES, WINDOWS, BRIDGES & OTHER CROSSINGS

Philippa Lewis

To Romy & Max,
grandchildren of the author and the illustrator

The drawings on pages 22, 27, 33 (bottom), 35, 36, 37, 40 (right), 41
(top & bottom right), 43 (bottom left), 44, 46 are by Miles Thistlethwaite
(www. milesthistlethwaite.com).

I have found most of the old woodcuts and engravings by wandering
through the wonderful open shelves of the London Library in
St. James's Square, London SW1.

Thanks to John Martineau for the sheer enjoyment in producing this book.

Victorian explorers in the mouth of a cave, portal between the dark rocky
subterranean realm and the world of light and life above.

CONTENTS

Introduction	1
Portals	2
Natural Portals	4
Thresholds	6
Gateways	8
Passages and Tunnels	10
Doors	12
Locks and Keys	14
Stairs	16
Slits and Small Windows	18
Large Windows	20
Chinese Windows	22
Screen Walls	24
Dry Stone Walls	26
Fencing for Humans	28
Fences for Animals	30
Gates for Animals	32
Stiles Over	34
Stiles Through	36
Tollgates and Turnstiles	38
Kissing Gates	40
Watery Gates	42
Crossings	44
Stepping Stones	46
Temporary Bridges	48
Bridges	50
Magical Portals	52
Gates to Other Worlds	54
Contemplative Gates	56
The Rainbow	58

Title plate to a 1632 English edition of Montaigne's Essays, engraved by Martin Droeshout, showing the book as an entrance or portal on to a world of ideas.

Introduction

THE WORD "PORTAL" derives from the Latin *porta*, which means gate, specifically a city gate, but also any place of ingress and egress.

A portal is generally an optimistic thing, both literally and metaphorically. A portal can be a door, window, arch, gateway, or just a gap in the hedge. The word encapsulates the idea of passing through, to a new opportunity, to making progress or moving forward, to entering fresh new worlds. Fifteenth- and sixteenth-century books often used the architectural motif of an elaborate doorway as a frontispiece, a visualization of the opening into the book (*see opposite*). Similarly a proscenium arch in the theater creates a "window" that separates the audience from the play which comes to life as the curtain rises.

Its companion word—"threshold"—describes the point on the ground under the portal, and can have the sense of being a springboard or a taking-off point: the threshold of success, the threshold of maturity.

As ever there is a flipside, doors and gates can lead into frightening and bad places: prison gates are entered through, but it may be some time before they can be passed again. There may be doubt as to whether a checkpoint can be crossed.

In a more abstract sense the word "portal" was adopted in the late 1990s, in the days when the Internet was called the information superhighway, as a name for a website which brought together information from diverse sources and gave them a consistent look and feel. Therefore through a portal came access to a multitude of other websites, and thus knowledge, like branches and twigs growing from the trunk of a tree.

Portals
transported

The Latin words for "carry" *portare*, "gateway" or "portal" *porta*, and "haven" *portus*, all have the same root, which gives the English words deriving from them an affinity. Thus by carrying our passports we can be transported elsewhere. We travel on ships and boats boarded at a port, or we fly off in an airplane that has touched down at an airport.

Ports, airports, and railway stations are all points of arrival and departure; as such they are often described as "gateways"—to the Continent, to the South, East, North, or West. In the architecture of railway stations this meaning has frequently been emphasized by grandly ornamented arched entrances or arches (*see opposite*).

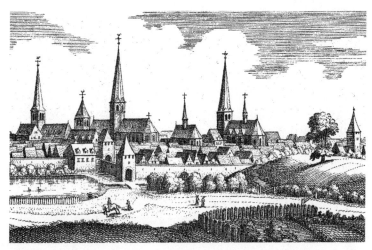

The old gateway into Essen, over a bridge and through two guardhouses; Matthaus Merian, 1647.

Transported by rail: Berlin Friedrichstrasse station.

Transported by sea: the Swedish port of Stockholm.

Transported by theater, cinema, or television.

Transported by air, on a plane, via an airport.

Transported by words in a book.

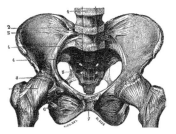
Transported to life through the birth canal.

Natural Portals
harbors and canyons

Landforms often dictate how man moves around the planet. Valleys, gorges, and canyons provide natural ways or passes by which to negotiate mountain ranges; estuaries or inlets along a rocky shoreline form sheltered harbors. "Port" (deriving from latin *portus*) means refuge as well as harbor; thus "a port in any storm." Humans have chosen to make settlements beside such places. The three-mile-long channel that runs from San Francisco out to the Pacific Ocean is named the Golden Gate.

Caves are another kind of natural portal, into a space protected from the elements (*see below*). Humans lived in caves for thousands of years.

Land is shaped by both constructive and destructive forces. Earthquakes and volcanic eruptions construct it upwards, while ice, water, tide, wind, or gravity erode and destroy. As glaciers and water will always find the path of least resistance, so these have naturally also become the easiest way for man.

Fingal's Cave, Staffa, in 1772.

Roger Rains House, Peak Cavern, 1809.

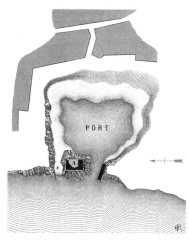

The ancient port of Jableh, Syria.

Mountain pass into the Sierra Morena, Spain.

Water bursts through rocks into pools below at Pistyll Cain, Snowdonia.

THRESHOLDS
liminal boundaries

A threshold provides a demarcation between the outside open world and an inside, protected, covered world, also a division between the outside world of public space and private space, into which admission is granted through the goodwill of the inhabitant (*below, center*).

A doorway is the entrance to a building, whether domestic, state, religious, educational, theatrical, or mercantile, and status is often expressed through its decoration, via classical pediments and pillars (*as below, left*) or heraldic escutcheons. Pairs of figures, maybe mythical beasts or lions, often stand as symbolic guards either side of important doorways. Imperial Chinese buildings were protected by pairs of male and female lions (also known as Dogs of Fo, or Lions of Fo). Ancient Assyrian guardian figures combined eagle wings, a human head, and lion or bull bodies. A pineapple ornament over a doorway is a traditional American sign of welcome, and projecting porches also offer shelter to the visitor at the door, both at the front and also at the back door, which exists for the mundane and the ordinary, such as goods and rubbish.

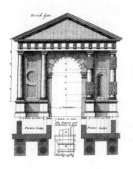  

Entrance to the Rock Temple of Ramesses II at Abu Simbel, southern Egypt.

One of the ancient gateways into Beijing, China, late 19th century. None surive today.

12th century Norman/Romanesque doorway to Dalmeny church, near Queensferry, Midlothian, Scotland.

Entrance to the medieval banqueting hall, at Haddon Hall, Bakewell, Derbyshire; lithograph by Samuel Rayner.

GATEWAYS
small and secret or public and impressive

A gate can provide a way through into something as large as a city (*previous page*) or as small as a garden (*opposite*). In both cases they can be closed as a barrier or opened in welcome. King Nebuchadnezzar II built in 575 BC the Ishtar Gate through the walls of Babylon, one of the Seven Wonders of the ancient world, a cedar gate set within a huge structure depicting aurochs and lions (*below*). A gate is the first impression to what lies behind and its design will often indicate its purpose.

Medieval towns and cities were encircled by walls for security; the town or city gate presented as a fortress or castle, of heavy stone, battlemented, castellated, and machicolated. The gates, which hinged open, or portcullis, which was raised up, were closed at night and times of danger. The word "Bar" (as in Temple Bar) stood for barrier, an alternative to a gate. Gates proclaim status: at the main entrance to a palace or country house its Brobdingnagian proportion was intended to make the entrant feel aware of the owner's superiority. Classical rustication, heraldic animals, crests, lion masks, trophies of arms and armor were chosen ornamentation for power and heredity on the gate piers or archway. By the 17th century the skill of the blacksmith was such that decorative iron gates of matching size and grandeur dazzled (*below*).

Above: Garden gate and fence at Avebury Manor, Avebury, Wiltshire.
Below: an idealized walled city, depicted by Bartolomeo Del Bene in 1609.

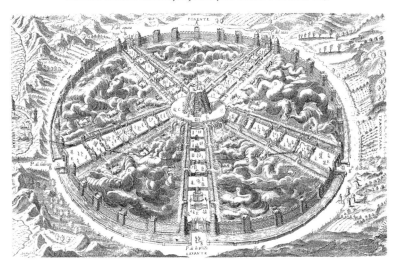

Passages and Tunnels
ways through and under

The word "passage" implies something tight and small, a narrow space between buildings, walls, or hedges that gives a route through. An underground passage is a tunnel: at its most magnificent it can go through a mountain, under a river, or under the English Channel.

Passages and tunnels can be secret, for stealthy travel, or just out of sight (*like the Parisian sewers, below*). Hidden entrances may involve a fake wall, a piece of furniture or trapdoor, or a concealed opening mechanism. Secret passages have existed for millennia: the Egyptians built them leading into burial chambers to confound grave robbers; ever since they have been used as escape routes, smuggling routes, and hiding places. They may also be ways of entering a building unseen.

Passages are sometimes covered ways: in warfare these provide shelter from the enemy (*opposite*), but they can also work like the "zingyan" of Mandalay, which protect Buddhist priests when they walk in the rain.

Entrance portal to subterranean levels discovered during excavations at Kouyunjik, Nineveh, 1852.

Various styles of temporary tunnels assist during the attack of a castle's curtain wall.

Entrance to a gold and silver mine in Colorado, drawn by Frenzeny & Tavernier in 1874.

A medieval sally port, from which sorties could be safely made by besieged/defending forces.

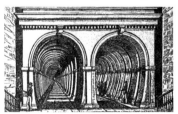

Entrance to the Thames Tunnel, built by Isambard Kingdom Brunel in 1843.

DOORS
open or closed

A closed door can express varying degrees of fortification. A medieval door of stout oak was studded with nail heads and long strap hinges, whereas a Victorian door was often partially glazed with stained glass depicting birds and flowers. The cynosure of an exterior, a door, expresses the style of the occupant.

To get a door opened, attention can be sought by beating with a fist and shouting. Louder and better is a knocker with the clang of metal on metal. A knocker might represent a grotesque animal or a fashionable ornament. A sanctuary knocker on a cathedral door, such as at Durham Cathdral (*center, below*), could be rapped to obtain 37 days of safety and shelter within. Originally, postmen knocked to deliver mail, until the advent of the Penny Post and letterboxes that allowed for a silent delivery. Bells were sometimes rung to appeal for entry, made louder with the arrival of electricity.

The threshold is the line crossed once a door is open; it also has the nonliteral meaning of "start" or "beginning." Two historical relics attached to doorways that relate to entering a house are the metal snuffers, into which the link boy thrust his torch to extinguish it, and iron boot scrapers, to clean the mud off boots and shoes in the days of rough roads and no pavements.

Above: a door fallen off its hinges no longer shuts. Below: Design for a studded and ornamented door, Paris. c. 1800.

Above: doorway to the Raphael Loggia at the Vatican. Below: iron hinges typically found in medieval cathedrals and castles.

LOCKS AND KEYS
only the chosen may pass

Possessing a key to a door signifies ownership or trust. Prior to the invention of the lock and key a door could only be secured from the inside with bolts and bars, or tied shut with complicated rope knots, which merely revealed if someone had tried to break in. Locks enable a place to be left empty, but secure.

The ancient Egyptians devised wooden locks, bolts with pins that fell into place holding it firm, which were only released by pushing a wooden key through the center of the bolt to raise them. The Chinese had portable padlocks as early as 1000 BC (mechanical *rocs*). The Romans introduced stronger metal locks with keys that were small enough to carry around. A warded lock is the most basic; it has a set of obstructions that shift only when the right key is inserted. Yet, despite multiple key designs and even fake keyholes, a nimble-fingered operator with a skeleton key could open most doors until innovators such as Joseph Bramah, Linus Yale, and Jeremiah Chubb invented more complex mechanisms integral to the door and its frame.

Secret watchwords and passwords have long been spoken when challenged by a sentry to gain admittance to an encampment, castle, or sensitive area, and are much in use again today. A watchword can also be a phrase that embodies the guiding principle. The modern descendant is the number pad, or a swipe card.

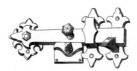 

MORTICE LOCKS AND LATCHES

No. Y 2016.

Size, ... 5 ins.
No. Y 2016.—Japanned Case, Iron Bolts... 19/6 per doz.
  „ Y 2017 B.—   „   „   Strong Brass Bolts, 25/6 „

No. Y 2023.

Japanned Steel Case, Round End, Narrow or Centre Bit, Mortice Lock, Brass Bolts.
Sizes, ... 5 6 ins.
No. Y 2023, ... 21/ 24/ per doz.
  „ Y 2028.—Half Rebated, ... 36/ 42/ „

No. Y 2021.

Strong Bright Steel Case, Round End, Steel Follower, Lever Action, Brass Bushed.
Sizes, ... 5 6 ins.
No. Y 2021, ... 37/6 41/3 per doz.
  „ Y 2022.—Half Rebated, ... 49/6 52/6 „

No. Y 2141.

No. Y 2141.—Japanned Case, Brass Roller Bolt ; size, 5 ins., ... 25/6 per doz.
  „ Y 2024.—Best Quality, Lever, Brass Roller Bolt, very Strong; size, 6 ins., 45/ „

No. Y 2024 B.

Bright Steel Case, Full Bushed, Two-lever, Scotch Spring, Palace Motion, Reversible Brass Bolts.
Size, ... 6 ins.
PRICE, ... 51/ per doz.
This Lock is specially recommended for School Board work.

No. Y 2061.

Extra Strong Square-end Mortice Lock, ⅝ in. thick, fully Bushed, Steel Follower, heavy Brass Bolts.
Size, ... 6 ins.
PRICE, Two-lever, ... 6/9 each.
  „ Four-lever, ... 9/ „

DRAWBACK, GATE, AND STOCK LOCKS

No. Y 2012 Fine Plate Lock.

Fine Plate or Wood Stock Lock, suitable for sheds or outside buildings.
Sizes, ... 6 7 8 9 ins.
PRICES, ... 15/ 15/9 18/9 26/3 per doz.

No. Y 2013 Best Strong Plate Lock.

Strong Plate or Wood Stock Lock, Brass Bushed, Steel Bound Edges, suitable for coach-house doors.
Sizes, ... 7 8 9 10 12 ins.
PRICES, ... 30/ 33/ 36/ 39/ 48/ per doz.

No. Y 2013 B Gothic Plate Lock.

Superior Finish Plate Lock, Varnished Oak Case, Fully Bushed, Brass Wards, Gothic Plated, Gothic Key, suitable for ecclesiastical work.
Sizes, ... 8 10 12 ins.
Extra Strong, ... 6/ 10/6 15/ each.

No. Y 2036 Drawback Lock.

Strong Japanned Drawback Lock, Fine Ward, Unbushed.
Sizes, ... 7 8 9 ins.
PRICES, ... 36/ 42/ 57/ per doz.

No. Y 2042 Gate Lock.

No. Y 2034 Drawback Lock.

Drawback Lock and Night Latch combined, very strong, handsomely bronzed.
Two Nickel-plated Keys, size, 6 ins., 58/6 per doz.
Extra Keys always in stock, 1/ each.

No. Y 2041 Gate Lock.

No. Y 2037 Drawback Lock.

Drawback Lock, Extra Strong, Covered Plate, Double Staple, Fully Bushed, with Brass Slide Catch.
Sizes, ... 7 8 9 10 ins.
PRICES, ... 48/ 51/ 67/6 90/ per doz.

No. Y 2094 Gate Lock.

STAIRS
over and under, ways up and down

Human feet work better on the flat, rather than at an angle. Since gods of nearly all cultures were believed to inhabit the skies, the route there was usually stepped. It is an acceptable theory that the monumental stone Step Pyramid of Djoser, designed by Imhotep in the 27th century BC, was not only a tomb but intended to facilitate an afterlife in heaven. The high ziggurat temples of the Sumerians and Babylonians had access by steps only for the priests. Mayan temples too were step-form (*below*). In Genesis, Jacob dreams of the way to heaven as by ladder with angels ascending and descending (*see page 55*).

Any desire line that involves a hill climb is likely to evolve into steps and staircases: down a cliff to a beach, up to a monument. Architects have long realized that to approach a building by flight of steps gives it gravitas. A straight staircase with no breaks is daunting; landings make pauses for breath and safety, turns and bends allow for flourishes.

The first working escalator, a moving staircase, was installed at Coney Island, New York, in 1896 (*below*). A giant outdoor escalator built in 2011 in Medellin, Columbia, has transformed the lives of the inhabitants of Comuna 13, a shanty town on the hills.

 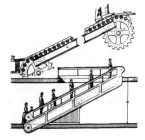

A wooden staircase, with two turns.

External staircase at Chateau-Gaillard.

Spiral staircase construction, 1688.

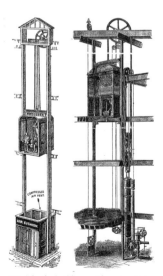

Otis lifts, for buildings with many stories.

SLITS AND SMALL WINDOWS
for light, eyes, and arrows

The English word "window" derives from the old Norse word *vindauga* or "wind-eye," an expression giving a hint to the fact that the earliest small windows, prior to the development of chimneys, drew out smoke from the central fire. Thus a window was a hole in the wall for ventilation: letting out bad air and letting in the good and allowing beams of light into an enclosed dark space. Wooden shutters provided security, warmth, and protection against cold, wind, and rain. A leap forward in comfort came when the problem was solved for sealing windows while preserving the light. The Romans first put glass in a window frame, but for centuries translucency was achieved with a ragbag of methods: translucent marble, mica, oiled cloth, horn, matting, etc. So precious was early glass that small casement windows (with a wooden or metal frame around the glass) were removable and considered a personal possession.

For defensive purposes, it was only through a slit opening, too small to enter through, that the outside world and any potential enemy could be observed in safely. By splaying the internal face of the wall the field of vision was widened. A long bow could be fired through a vertical slit, a cross bow needed a width-ways opening as well.

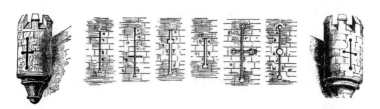

Opposite page: defensive balistraria, through which bows and crossbows could be fired. Above top left: example from Micklegate Bar, York. Bottom right: exterior crenelles with wooden hanging shutters. Top right: secure windows in Irish round towers. Bottom left: a gunport for a cannon.

Large Windows
for views and letting in light

As techniques in glass-making developed it could be produced in ever larger expanses, transforming the size of windows and thereby interiors by admitting ever more light; and with more light, more could be accomplished. Conversely, looking out from the interior, windows revealed ever more of the world outside. The window became a dominant feature of an exterior wall and architects devised windows of all shapes and sizes: oriels, bows, bays, rounded, oval, dormers, skylights, French windows, and picture windows. In Philip Johnson's Glass House, built in Connecticut in 1949, the walls become entirely glass. At night the effect of a window is reversed; the view—a lit interior—is visible only from the outside. Moths, oblivious of the barrier, beat against the glass to reach the light.

We want to look out of houses, but tend to fear people looking in, so we cover windows with blinds and curtains at night. However, we are encouraged to press our faces close to shop windows and consider the riches within—window shopping (*see New York storefront, 1865, below*).

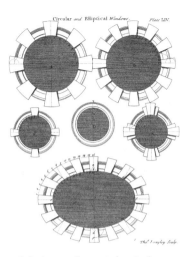

Left: A woman leans out of a pair of open windows, Paris, 1880s. Above: round and oval windows, from a pattern book of 1739.

Above: shutters, blinds, and curtains allow control over light coming through windows. Right: a 17th century engraver in his workshop.

CHINESE WINDOWS
for framing landscapes

The Chinese saw windows as what the Chinese Ming period scholar gardener Li Yu called opportunities for "unintentional painting." The window opening, which could be circular, hexagonal, gourd, petal, or fan-shaped, became frame to a carefully composed garden or natural landscape feature when looked through from within.

The interior therefore has scenery "borrowed" from the outside, often looking to a landmark such as a pagoda. Sometimes the openings were traceried with features such as trees and rocks (*below left*) or were bordered with geometric fret patterns (*below right*) which further elaborated the view through and varied the perception of distance. These were particularly used in garden pavilions and covered ways. Fan-shaped windows were also known to have been used on pleasure boats, so the passing landscape was seen as a sequence of fan "paintings."

A moon gate (*see opposite*) is a circular opening in a wall in Chinese gardens. Each opening is a reminder to pause and reflect before walking through to another part of the garden. Both the lead up to the moongate and the view through it are careful compositions.

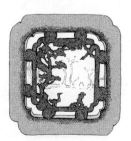

Left: two pictures of Chinese moon gates, by Bertha Lum, c. 1936. Above: octagonal windows in a Chinese courtyard. Below: hexagonal doorway connecting a Chinese exterior garden to an inner courtyard.

Facing page: elaborate tree & rock, and geometrical tracery, create frames for important views of the garden through Chinese windows.

Screen Walls
only the eye may pass

The most basic openwork pattern is a diagonal or vertical/horizontal lattice or trellis. It is a boundary that protects and encloses a garden space, but fragile since it is simply made from wood. As the interstices provide a structure for climbing plants, so it can become a living wall of greenery and flowers. Although sometimes constructed of masonry the medieval *hortus conclusus* (enclosed garden) is usually depicted as being constructed of trellis and can be emblematic of the Virgin Mary, whose symbol, the rose, may be entwined on it.

Wooden or plaster screens perforated with geometric patterns (Mashrabiya/Mousharabi/Jali) are a traditional element of Islamic architecture, used since the Middle Ages. They filter out heat from the sun while allowing a flow of air in the building. They also permit a discreet peep through to the outside world while providing privacy essential for women, thus can equally be found in the women's quarters of Moghul palaces or in a Sicilian church, where the nuns of a closed order sing and pray behind a grille.

Above: a trellis garden, from Hypnerotomachia Poliphili, *Italy, 1499. Left: 16th century sandstone and wooden Jali screens, from Agra, India. Opposite page: decorative Victorian iron window grille; carved window from the Mosque of Ibn Tulun, Cairo, Egypt; screen from Tomb of Humayun, Delhi, India.*

DRY STONE WALLS
mice, insects, and birds may pass

A wall built without mortar is ageless. Piling one stone on another may be a primitive instinct but it is also a skilled technique, as exhibited most dramatically at Great Zimbabwe built by the Shona chieftans between 11th and 15th centuries, with walls, probably an expression of power and status, reaching 36 ft in height and 820 ft in length.

On rocky land with thin soil such walls evolved naturally as land was cleared of rocks and boulders for cultivation. On hilly landscapes they held terraces firm for growing vines and olives. The weight of stone means that it is never carried far, so a wall reflects the area's geology. A wall is always been built from top to bottom so stone is never carried uphill, but rolled down.

A dry stone wall is porous; water runs through the gaps and it stays firm. Interstices provide a highway safe from predators for small rodents, birds, invertebrates, and insects to which the wall is no barrier. Low square openings were sometimes made to allow sheep or watercourses through—British regional terms for these include cripple, snout, or lunky holes. Alternating large and small capstones are called cock and hen, or buck and doe.

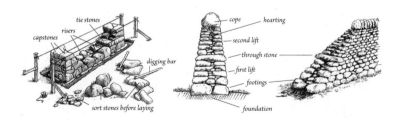

Above: various styles of English dry stone walls, built without mortar: i. boulder wall; ii. bolthole or smoot for animals; iii. vertical stone wall; iv: herringbone wall; v. & vi. with vertical capstones.

Fencing for Humans
pickets and pallisades

Fences are a tangible expression of a boundary. They divide private defensible space from public space, or they might surround something dangerous—a deterrent.

Fences are generally made from wood or iron. The most primitive and earliest, used from the ancient Greeks through to American colonists, is a palisade or stockade (*see Maori and U.S. examples below*). It is fast fortification: upright wooden stakes driven into the earth and close-packed, their tops sharpened into a point; effective, but vulnerable to fire. The wooden picket fence (*p. 31*) is a polite stockade, the uprights are spaced apart allowing views and small animals through. The tops might be rounded, but, neatly painted, it nevertheless expresses control and possession of everything enclosed within it.

An iron fence, or railing, is harder, longer lasting, and less friendly, and became popular once the casting process was technically mastered in the late 18th century. Finials on the uprights frequently imitated weapons: arrows, bayonets, axe-heads, fasces, even mace, sending a painfully sharp message to the casual intruder.

Mending your fences is good practice, leading to clarity.

Above: pallisaded village, Africa. Left: cast iron fencing and finials. Below: picket fence, 1485 woodcut. Opposite page: Maori stockade, drawn by Ethel Richardson, c. 1820; Civil War town stockade, New York State, c. 1863.

Fences for Animals
hedges and ha-has

Julius Caesar noticed hedges in Flanders in 55 BC. Saplings were cut and bent and then inserted with briars to make an impenetrable natural wall, which is how hedges have been laid ever since. A hedge coerces nature into making a boundary, corralling stock, and incidentally providing fuel and food to men, food and shelter to small beasts; barrier as ecosystem. Pioneers in America devised zig zag, worm, or snake fences which could be erected on hard and rocky ground, a self-supporting system of interlocking split rails that needed no nails or holes driven into the ground.

With wood in short supply, fences had to be made of wire; animals could push over smooth wire, but not barbed wire, man's answer to a thorny stock proof hedge, quicker and cheaper. Devised in America, the first patent was granted to Lucien B. Smith of Kent, Ohio, in 1867 and transformed the cattle industry in the America by enabling thousands of acres to be cheaply enclosed.

So-called "park fencing" is the gentrification of animal fencing, and used within sight of the house. An alternative to this was the ha-ha, where a wall and ditch provided a seemingly invisible barrier between grazing animals and garden grounds (*below, right*).

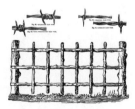  

Above: 15th century print of a garden surrounded by a woven wattle fence.   Below left: a hedgelayer plies his trade, 1945 etching by Stanley Anderson.   Below and opposite: various rustic fencing systems, some involving barbed wire.   Below right: a floating fence for a forest stream, allows logs to pass under.

Gates for Animals
and for children to swing on

Herding is the age-old practice of keeping flocks of animals intact—thus shepherd, swineherd, goatherd, goose girl—but as farming evolved it became customary to pen animals within a field to graze and feed. A field needs a gate. Field gates must be strong and light and wide enough to allow a rush of animals through at the same time. They must be low enough to prevent animals squeezing underneath and high enough to prevent them jumping over. Utilitarian, they must be economical to manufacture. The traditional solution is a varying combination of horizontal bars strengthened by perpendicular or diagonal struts, a five-bar gate, the design changing from region to region. Seasoned oak was considered the best British material. Where wood was scarce the gateposts might be of stone slab or granite. The hinges and latches were traditionally made by the blacksmith, of patterns as various as the men who forged them. Some gates self-close with a rising hinge or a weight hung on it, some swing open both ways. Fastenings need to be easily opened by remaining firmly fixed.

A cattle grid will generally prevent animals from straying out of their field, useful where there is a track through; but as it has been known for cows to become adept at crossing the bars despite their cloven hoofs, it is not a guaranteed animal-proof solution.

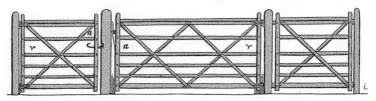

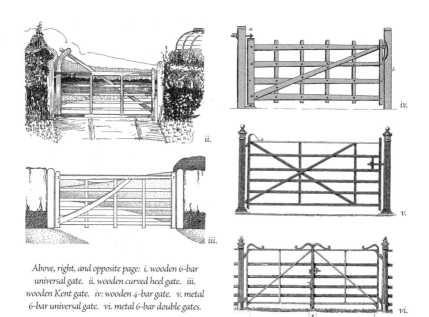

Above, right, and opposite page: i. wooden 6-bar universal gate. ii. wooden curved heel gate. iii. wooden Kent gate. iv: wooden 4-bar gate. v. metal 6-bar universal gate. vi. metal 6-bar double gates.

A cattle grid is a gate designed on the same principle as a stile, allowing passage but keeping animals in, and often used on roadways. It can be navigated by wheels and flat human feet, but not by hooves.

Stiles Over
permitting pedestrian passage

The function of a stile is to simplify the way over a fence, hedge, bank, or wall for pedestrians in single file, but keep the livestock penned. They are as old as enclosed fields when rights-of-way and footpaths to water, wood, or village became blocked.

Steps were needed to get over walls, particularly dry stone with a tendency to tumble; these might be in the manner of a staircase built alongside or across a wall. They might be large stones projecting from the wall itself. A single slab of stone might be inserted in the wall to make it easier to climb, either vertically (*see examples from the Cotswolds drawn by William Simmonds in 1939, below*), or protruding or sometimes with a hole to push a foot through. A wooden ladder placed like a hairpin over the wall is the easiest construction but not necessarily as long-lasting. A wooden stile will fill the gap in a hedge or provide a way through a fence. How the step footholds are arranged is a matter for the maker to decide, as is the inclusion of a retractable flap for a dog, or an upright pole for extra ease of climbing.

 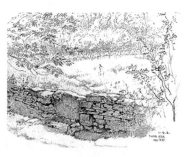

Ladder stile, common in Northern Ireland, Snowdonia, and the Lake District. Good for getting over and protecting high walls.

Step stile, made from protruding stones inserted at the time of a wall's construction. Need a heavy wall, or walkers will destabilize them.

Wooden stile with two crossed steps, for climbing barbed wire or wooden fences. One of the most common and easy-to-use country stiles.

Concrete step stile, northern France, essentially a staircase in a hedge or fence. Too steep for larger animals, though porous to goats.

Stiles Through
narrowing the access

If not over, then through: another way to bar cows, horses, and sheep but let man slip through is through a squeezer or vee stile. This is an opening so narrow that it is only wide enough for a man who might have to pass through sideways. This style of stile can be built in stone (*see example from Goldhanger, Essex, below right*), wood (*below left*), or metal (*as in the example from Owlpen, Gloucester, opposite lower right*). A simple squeezer stile can be built out of a curved or J-shaped branch, sawn or split down the middle and reflected into two sides.

A zig-zag passage was also used to baffle the quadrupeds. A stile of falling bars, or clapper stile, consists of bars that are light at one end, heavy at the other. The bars are fixed on the post with pivots, so when the lighter end is pushed down they fall to the ground and can be easily stepped over (*shown on page 29*).

In the spirit of improvement, various mechanical devices were applied to the principle of the stile. A draper called Thomas Lyne invented one with upright posts geared at ground level that fall sideways when a knob is pushed/pulled.

Above: cludge stile, made from odds and ends. Note the horizontal bars, which bridge the two round posts, transferring the tension in the wire between them; also the useful hand pole. The hole beneath is big enough for a dog. The design has also incorporated a tree stump.

Above: bridge stile over a ditch or stream, allows humans to cross, but not animals.

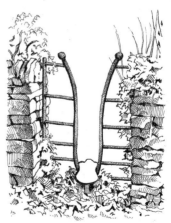

Above: metal squeezer stile, with step polished from years of use.

Tollgates and Turnstiles
one way and one at a time

At a tollgate you exchange passage for payment, at city limits, bridges, and on turnpike roads. A turnstile is a member of the stile family, but one whose original simplicity developed extra ramifications and became a subject for many patents. Also called a baffle gate, the simplest form of turnstile was a revolving bar that blocked the passage for four-legged animals, but allowed two-legged ones to push through.

The turnstile evolved into a mechanism that allowed passage in a single file and simultaneously counted the numbers of those going through. Further sophistication involved a braking system on the revolving bars that was released by inserting a coin, token, or ticket in a slot. This kind of automatic barrier is today used for gaining paid entry to transport systems, lavatories, swimming pools, museums, games stadia, playgrounds, and amusement parks.

A turnstile can be used as a security measure by allowing only those with a pass or code through it. Some will let people through in both directions, some are one-way only. The gentlest turnstile is waist high, but they also exist at full height, thus preventing anyone vaulting over without being counted. These are sometimes called "iron maidens."

Country tollgate beside a cutting through a hill, the toll paying for the upkeep of the road.

Turnstile and toll collector with broom, near St. Paul's Cathedral, London, c. 1830.

Above: Waterloo Bridge turnstile and tollbooth, London, c. 1880, with pedestrian barriers similar to the ones used today on the London Underground. Opposite: rural turnpike; Putney Bridge tollhouse.

Kissing Gates
and revolving doors

Two relations of the turnstile are the kissing gate and the revolving door. The kissing gate flaps back and forth, always maintaining a barrier, but allowing passage through. Perfect for footpaths through fields full of stock.

Revolving doors are sophisticated turnstiles, but their purpose is different: unlike a hinged door, they provide a constant entrance and exit to a building without letting rain, snow, wind, or dust blow in from outside. The perpetual air lock eliminates draft and noise. It has three or four partitions that revolve in a circle, fitting the circular entrance snugly. People can simultaneously come in and go out.

H. Bockhacker of Berlin was granted a patent in Germany in 1881 for a "door without an air draft" and Theophilus van Kannel of Philadelphia one in 1888, calling it a "Storm-Door Structure." But it was another ten years before the world's first wooden revolving door was installed in 1899 at Rector's restaurant in Times Square, New York.

In right-hand driving countries doors typically revolve counter-clockwise, in left-hand driving countries clockwise.

Above, and opposite page: kissing gates, in wood and iron. Kissing gates allow one person through at a time, but their design means that large animals cannot pass.

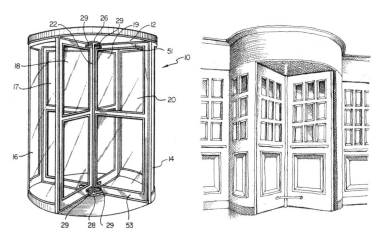

Above: revolving doors, in metal (left) and wood (right). Revolving doors allow one person through at a time, and have the additional advantage of preventing loss or addition of cold or hot air to the interior.

WATERY GATES
sluice, flood, and lock gates

Sluice gates control the flow of water on rivers and canals, to prevent flooding or to maintain a flow to a watermill. When lowered, they block the passage of water; when lifted the flow is restored. Simple sluice gates are hand-operated (rack and pinion, worm driven or chain pulley system); others are hydraulically or electrically powered.

Cutting a passage through land to speed travel and transport goods creates a canal, but since land is never flat, systems involving gates to change water levels were devised. The earliest system was a Flash Lock which only worked going downwards by temporarily damming the water and then releasing it—the boat traveling on the surge. The pound lock improved on this, enabling a boat to travel up or down hill. Used by the Chinese in the 10th century and in the Low Countries from the late 14th it involved a fixed chamber (pound) with a guillotine-type gate at either end. A drawing by Leonardo da Vinci (*below, left, from Il Codice Atlantico*) illustrates his invention, the Mitre Gate, which he used on a lock in Milan in 1497. Its V-shape was held together by water pressure and a small wicket gate at low level accelerated the system: a design that has never been improved on.

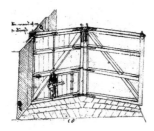

Teddington locks, 1830. Locks enable boats to climb hills by entering chambers which are then sealed and flooded from upstream, raising their levels. The Caen Hill locks near Devizes, Wilts, consist of 29 individual locks which climb 237 ft over 2 miles.

Above: sluice gate, Somerset. Where water levels are important, for example in canals, or in heavily drained areas of land, and near watermills, sluices, and weirs (opposite right), they allow fine control of water flow and are opened after heavy rainfall to prevent flooding, or to allow the controlled flooding of water meadows.

Above: Traitors' Gate, the water entrance from the Thames into the Tower of London. Many who passed through it never emerged again.

CROSSINGS
rights of way

A ferry boat is a path over water. The most famous ferryman is Charon who, in Greek mythology, transported the souls of the newly dead across the rivers Styx or Acheron to Hades, the underworld (*see page 54*). A coin placed in their mouth paid the fare. Ferry crossings (*opposite, top*) are an alternative to bridges. A well-trodden route over the centuries often becomes a *de facto* right of way or a permissive path. It might lead to a ford (*opposite, lower*), a shallow spot where it is possible to cross a river. Likewise, footpaths (pedestrians only) and bridleways (riders of horses and bicycles) give free passage over private land.

Some crossings make one kind of traffic stop to allow others to safely pass, such as a railway crossing (*below*), traffic lights, or a zebra crossing. In all these cases flashing lights are used as a warning symbol.

Ferry crossing over the Upper Thames, using a barge and a fixed line. The box-like arrangement in the left foreground is a landing stage for use when the river has risen to a higher level.

Ford across a stream. Ancient fords tend to be found at places where the river bed is naturally rocky and does not collect mud, even after floods. Newer fords can involve laying a stone path.

Stepping Stones
precarious ways over water

Water is often a barrier to travel, but there has always been a desire to cross to the other side and humans and animals have always sought out the shallowest and safest part to cross—the ford (*see too previous page*). Fords are revealed in place names, important in a pre-map age: e.g. Brentford, Harbertonford, Stratford, Hartford (where stags crossed).

Wading through a ford meant wet feet which travelers solved with stepping stones. They probably evolved slowly as good firm stones were found and added to the desired line. A flood can easily wash away stones or take them out of line, so stepping stones are by their nature impermanent or constantly changing.

Striding or jumping from stone to stone can be beset with pitfalls with the chance of wobbling, slipping, or losing balance; stepping stones are therefore often used as a metaphor for slow and careful progress, each step being an achievement, bringing the walker nearer a goal.

The combination of stone and water is beloved of garden designers and sculptors, the random roughness of stones replaced by perfect circles, like lily pads on water, on which, sadly, man cannot walk.

Various ways of crossing water and reaching the other side, engraved by Thomas Bewick [1753–1828], a native of Northumberland, including fords, stepping stones, poles, trees, stilts, rope bridges, ladders, and planks.

Temporary Bridges
pontoons and drawbridges

To invade Greece in 480 BC, the Persian Emperor Xerxes crossed the Hellespont on a line of boats lashed with flaxen cables, a precursor to the World War II Bailey bridge (*below*). Classic medieval castle design includes a moat (or at least a ditch) as an extra fortification. A refinement on an ordinary wooden bridge that could be abandoned in emergency was the drawbridge. The hinged bascule could be winched up from the gatehouse above the threshold, with the aid of counterweights, to make a barrier. A drawbridge was often combined with a portcullis, a strong grille of wood and metal, that could be rapidly lowered to bar entry. Some castles had two portcullises so the enemy could be trapped between them as they invaded. Once the bascules could be managed by steam power rather than manpower, the weight and length of each "leaf" could be considerable.

The route over the water was now interrupted to allow tall shipping along the waterway rather than to bamboozle the enemy. Perhaps the most famous is Tower Bridge (*see picture on p.50*).

Top left: double drawbridge at the Castle of Montargis, France. Above right: various drawbridges. Left: movable bridge, used for crossing the small arm of the River Abonia during the seige of the ancient city of Juliana, Burgundy, 522AD, from Eugène Viollet-le-Duc. Left: Bailey bridge, a lightweight prefabricated truss bridge, strong enough to carry tanks, widely used by British, Canadian, and American forces in World War II, invented by Sir Donald Bailey.

BRIDGES
ways over water or chasms

A tree falls across a river and makes a bridge: an accident of nature. A bridge might be as simple as a plank of wood for walking over a stream or the six-lane highway linking Qingdao on mainland China to the island of Huangdao that stretches for over 26 miles. Bridges cross chasms, lakes, swamps, and the sea; they have been built for pedestrians, packhorses, wagons, carts, coaches, cars, juggernauts, railway trains, and, on an aqueduct, canal boats. Bridges are structures of daring and imagination and crossing one requires having faith in its strength. Bridge-builders have constantly defied the naysayers.

The Romans showed the world how to build stone arches, enabling the creation of multiple spans to cross the widest river. The 1st century AD Pont du Gard has two tiers, carrying an aqueduct above a road.

An inverted version of an arched bridge is a suspension bridge, which carries the weight of the track via cables suspended from towers. The Inca created fragile suspension bridges across canyons from woven natural fibers renewed annually to ensure their safety. Such bridges of iron and steel still sway in the wind.

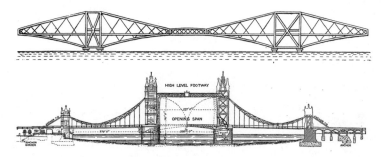

Above: Eltham Bridge, by Eltham Palace, Kent, in 1826, which still stands today. In this summer scene the river has all but dried up, but come the rains the river will return.

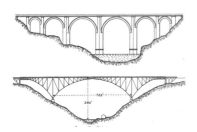

Left, upper: high arched bridge, with five columns supporting a road or waterway, first built by the Romans. Left, lower: Viaur Bridge, France, a 3-hinged steel truss arch. Opposite page, top: Forth Bridge, a cantilever railway bridge over the Firth of Forth, Scotland. Opposite page, lower: Tower Bridge, London, a bascule suspension bridge.

Above: Khaju Bridge, Isfahan, Iran. Built c. 1650 by the Safavid Shah Abbas II, it supports a covered passageway and central pavilion, as well as controlling the river flow as a weir with sluicegates.

MAGICAL PORTALS
in fiction and lore

Writers have long used imaginary portals in stories to enter fantasy worlds, for in fairy tale, magic progresses the narrative journey.

Only with the phrase "Open Sesame!" can Ali Baba in *Arabian Nights* open the blocked door to enter into the cave where the forty thieves have hidden the untold treasure. Lewis Carroll's Alice follows the White Rabbit down a rabbit-hole and falls softly into her *Adventures in Wonderland* that begin after she has drunk a magic potion allowing her to be small enough to pass through a 15-inch high door opened with a golden key. In the sequel she passes *Through the Looking-Glass* (below) to enter a fantastic world of live chess pieces and unicorns. The four children in C. S. Lewis's books enter Narnia, a land of talking animals and mythical creatures, through a wardrobe in the spare room of a country house and then through a painting on the wall.

Above: a fairy circle, portal to the otherworld, can be a circle of mushrooms, stones, or a raised tump as here, illustrated by Richard Doyle [1824–83]. Below: Castlerigg Stone Circle, Cumbria, 3200 BC.

GATES TO OTHER WORLDS
and the born again

In many cultures there are upper worlds where the gods live (Olympus, Heaven, Asgard) and a contrasting underworld for the dead, sinful, and dangerous (Hades, Hell, etc). Some were accessible to mortals, others not. The portals and routes to these other worlds are various.

The Valkyries carry the bravest slain warriors over the bridge to Valhalla (which has 540 doors). In Dante's poem *La Divina Commedia* the entrance down to hell is through a gate inscribed "Abandon all hope, you who enter here." In Greek mythology Charon carries the shades of the recently deceased across the river Styx (*below*).

In the journey made by Christian in John Bunyan's *Pilgrim's Progress* he passes through a wicket gate (a small pedestrian gate within a larger one) opened by the gatekeeper Goodwill letting him through to the straight and narrow path.

Jacob's ladder, between Heaven and Earth, appears to Jacob in a dream as he takes flight from his brother Esau in Genesis 28:10-19.

A lychgate (Dutch "lijk," German "leiche," Norse "lik" = dead), typical of English churchyards, for resting corpses prior to burial.

Christian, from John Bunyan's Pilgrim's Progress, is greeted by Goodwill at the wicket gate, much like St. Peter guarding the gate to Heaven.

Cerberus, the many-headed dog, guards the gates to the Underworld. Orpheus lulled Cerberus asleep with his music.

Contemplative Gates
labyrinths and altars

Spiritual journeys have one destination but many routes. The Romans made mosaics showing Theseus killing the Minotaur, as a reminder of the need for self control. Muslims pray toward Mecca, in a direction indicated by the mihrab in a mosque. Christians use an altar or cross.

The eye can be perceived as a portal through which one sees the world around but which also in reverse reveals an interior: hence the expression "The eye is the window to the soul." Cicero saw the face as a picture of the mind with the eyes as its interpreter. Eyes allow light to pass into us and fall onto sensitive exposed parts of the brain.

The Third Eye, inner eye, or mind's eye refers to the gate that leads to the interior being and spaces of higher consciousness. Thus in Hinduism the third eye of the god Siva is situated on the brow, and is the eye of spiritual wisdom and knowledge. He uses the third eye to see beyond the apparent and separate good from evil. Buddha's third eye is the eye of consciousness. The third eye is the gateway to visions, clairvoyance, and out-of-body experiences.

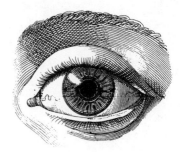

Theseus and Ariadne, outside the labyrinth. Many Roman labyrinth mosaics depict Theseus slaying the minotaur at their center, a reminder of the sacred duty to master the animal passions.

The mihrab (direction of Mecca), at the Suleymaniye Mosque, Istanbul, Turkey.

The angel Gabriel welcoming a pilgrim at the pearly gates of Heaven.

THE RAINBOW
gold at the end

The rainbow appears as a portal or bridge in many myths around the world. The Norse saw it as Bifrost, the burning bridge which connected Midgard (our world) with Asgard (the realm of the gods). In Judaism and Christianity it symbolizes the pact of peace between heaven and Earth. In Australian and South American legend it takes the form of a creator god, or rainbow serpent. Irish folklore has leprechauns hiding a pot of gold at the end of the rainbow. Today, the rainbow stands for peace, tolerance, respect for life, and diversity.

Rainbows are portals through which no one can physically pass, as they recede from the viewer as fast as they are approached. Like all portals they remind us of journeys to be made, doors to be opened, mysteries to be explored, and gold to be found.